Junk Food

WHERE'S THE SCIENCE HERE?

Junk Food
WHERE'S THE SCIENCE HERE?

VICKI COBB

Photographs by
Michael Gold

Millbrook Press   Minneapolis

For my granddaughter Jillian Cobb,
who likes all kinds of foods.

Author Acknowledgments
The author gratefully acknowledges the contributions
of the following people but takes full responsibility
for the accuracy of the text: Wendy Rappel, of the
Popcorn Board; Christine M. Dugan, Shawn M.
Bennet, and Judy Cooley, of Hershey Foods; Daryl
Thomas and Phil Bernas, of Herr Foods, Inc.; and Ron
Mancini, of Mother Myrick's, Manchester, Vermont.

Photographer Acknowledgments
Ed Herr and Suzy Stahl of Herr Foods, Inc.; Matthew
Cordovano; Timothy Hurley; Taylor Rossi; Alex Rossi;
Sophie Jennis; Karl Krause of Krause's Candy Store
in Saugerties, NY; Nicole Kurek; Rachel Nagler; Rose
Kurek; Maria Cordovano; Linda Gold; and Gabrielle
Gold.

Photographs by Michael Gold except for the
following: © SuperStock, Inc./SuperStock: Front
cover; © Loren Winters/Visuals Unlimited:
p. 10; © Jack Fields/Photo Researchers, Inc.: p. 22

Text copyright © 2006 by Vicki Cobb
Photographs copyright © 2006 by Michael Gold

Millbrook Press
A division of Lerner Publishing Group
241 First Avenue North
Minneapolis, Minnesota 55401 U.S.A.

Website address: www.lernerbooks.com

Library of Congress Cataloging-in-Publication Data
Cobb, Vicki.
Junk food / by Vicki Cobb.
p. cm. — (Where's the science here?)
ISBN-13: 978-0-7613-2773-8 (lib. bdg. : alk. paper)
ISBN-10: 0-7613-2773-8 (lib. bdg. : alk. paper)
1. Food industry and trade—Juvenile literature.
2. Junk food—Juvenile literature. I. Title.
TP370.3.C63 2006
664'.6—dc22   2004029821

Manufactured in the United States of America
1 2 3 4 5 6 – DP – 11 10 09 08 07 06

Contents

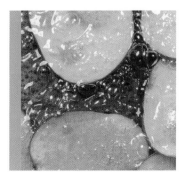

Why You Can't Eat Just One

Yum! Junk food tastes *sooooo* good! One piece of candy, one potato chip, one kernel of popcorn is simply not enough. The tastes of sweetness and saltiness are hot-wired into our brains at birth. Our tongues enjoy these tastes right from the start. Human breast milk is sweeter than cow's milk, so our earliest survival is linked to sweetness. Fat, such as the oil used to cook potato chips or the cocoa butter in chocolate, is particularly good at carrying delicious flavors. So taste is a big reason why most of us love junk food.

"Junk food" is slang for foods that don't have a lot of *nutrients*—the kinds of substances needed to keep you healthy.

Can you name seven kinds of junk food decorating this cake?

Fresh fruits and vegetables are essential to eating healthily.

It's especially important to develop healthy eating habits when you are young. You should eat a wide variety of foods because each food makes its own contribution to your overall nutrition. Most of the food you eat every day should include lean meat or poultry, low-fat dairy products, and complex carbohydrates, which are starches that are not refined like white flour and sugar. They include cereal, whole grain bread, beans, fruits, and vegetables. Complex carbohydrates should be just over half of the food you eat every day.

A diet of only junk food can make a person overweight and lead to other medical problems such as diabetes and heart disease. It's not hard to figure out which foods have low nutritional value if you learn to read the label on the package. The law requires food manufacturers to list the ingredients as well as certain measures of nutrition in their products. If the label on a package lists a form of sugar, fat, or salt as one of its first three ingredients, chances are good that it's some kind of junk food.

As long as you eat a healthy, balanced diet, many experts believe there's nothing wrong with a junk food treat every now and then. There are a lot of scientific ideas behind candy and chips. Most of these ideas are chemistry—the science of matter and how it changes. Other ideas come from the science of nutrition. Read this book and munch while you digest a little science.

Popcorn

Popcorn is the result of an explosion. Water inside a popcorn kernel changes into steam when the kernel is heated. The tough outside hull of the kernel acts like a watertight container, keeping the steam confined. Since the water is spread throughout the soft starch of the kernel, the expanding steam makes tiny bubbles in the hot starch. Pent-up steam builds up in pressure, putting more and more force on the hull until it can't take it anymore, and it ruptures. This foam cools quickly to become the firm white mass that you like to eat. That's the scientific explanation of what makes popcorn pop. But it wasn't the original explanation.

This multiple exposure shows a popcorn kernel in the act of popping. Notice how the kernel turns inside out as the starch expands.

Popcorn Facts

- Popcorn right off the cob is 16 to 19 percent water. It is dried to about 13 percent water for best popping.

- In order to explode, a kernel must be heated to about 450°F (232°C), at which point the pressure is about 135 pounds (81 kilograms), or 9 times atmospheric pressure. (This would mean that you'd be squeezed with 135 pounds on every square inch of your body. Normal atmospheric pressure is 15 pounds per square inch, and you are so accustomed to this that you don't even feel it.)

- There are two shapes for popped popcorn. "Butterfly," or "snowflake" (very irregular and pointy), is the kind you pop at home. And "mushroom" (round and smoother) is the kind used for caramel corn because it coats well.

Snowflake or butterfly popcorn

Mushroom popcorn

Dissect a Popcorn Package

Cut open a package of microwave popcorn.

- The outside layer of the bag is paper—the most inexpensive kind of packaging material.
- The inside of the bag is lined with a thin layer of the same kind of plastic that is used for soft-drink bottles. It is a material known to be nonreactive with foods.
- The section of the bag that is placed down in the oven contains a piece of metallized film called a *susceptor*. It is prepared by spraying plastic film with a fine layer of aluminum. The susceptor lies between the paper and the plastic lining. When microwaves strike the susceptor, the aluminum molecules give off heat energy, which melts the fat. The hot fat heats the popcorn in the bag the way oil heats popcorn over a stove.
- The corn is embedded in solid fat. When the fat gets hot, it distributes the heat to the kernels. The fat also contains the flavoring—and a lot of extra calories.

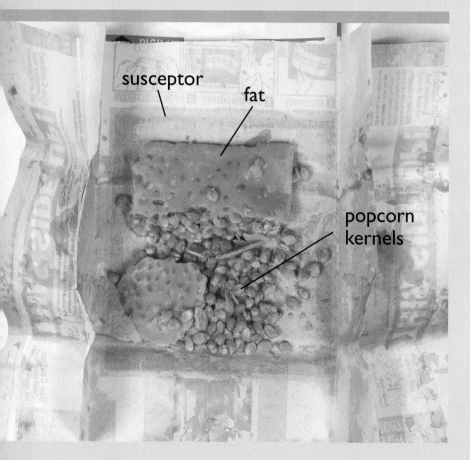

susceptor

fat

popcorn kernels

- Compare the fat in a regular bag of microwave popcorn to the fat in "lite" popcorn. Notice the difference in color. Feel the fats. Which melts first from the heat of your hand? Fats that are saturated, like butter and most animal fats, are solid at room temperature. Unsaturated fats are oils, which are liquid at room temperature. If you add hydrogen chemically to an unsaturated fat, it will change into a more saturated one and become a solid. Food manufacturers add just enough hydrogen to vegetable oils so that they are solid enough to hold the popcorn. Since they have less saturated fat in them they are called "lite."

- Do you need a prepared package to microwave popcorn? Try popping some plain popping corn. Pour about $1/4$ cup of popping corn in a microwave-safe bowl. Cover loosely by placing a plastic cover on top of the bowl. Microwave for four minutes, but be prepared to turn the oven off when the popping stops. Do you have more or fewer unpopped kernels compared with packaged microwave popcorn?

Popcorn was probably discovered by the Indians of North America and Mexico and is the first way people ate corn. Ears of popcorn that were more than 5,000 years old were discovered in the Bat Cave of New Mexico. The American Indians believed that little spirits lived in each kernel of corn. They thought the spirits grew angry when their houses were heated so they exploded in a temper tantrum—definitely not a scientific explanation!

Like all grains, corn was an important food because it could be dried so that it didn't spoil. It could be stored and eaten for many months after a harvest. Popcorn seeds are too hard to chew and difficult to grind into flour. Putting them over heat so that the kernels exploded into a soft white food was practical and delicious.

The Indians popped popcorn in clay pots over fires. We usually prepare it in bags in microwave ovens. This amazing technology would not be possible without the scientific study of light.

The light you see is a form of energy that travels in waves. You can't see light waves, but you have seen waves in water. One property of all waves is that they have a length that's measured from peak to peak or trough to trough. Each color of a rainbow—red, orange, yellow, green, blue, and violet—has its own wavelength. Red has a longer wavelength than orange, orange is longer than yellow, and so on. In addition, there is a whole range of energy that cannot be seen. Invisible infrared light and radio waves are longer than red. Ultraviolet rays and X-rays are shorter than violet. The invisible rays that caught the interest of the food industry are microwaves—longer than infrared but shorter than radio waves.

Microwaves are anywhere from a fraction of an inch to several feet long. They can penetrate plastic, paper, and glass. They have a powerful effect on water molecules. Normally, water molecules at room temperature are moving, tumbling over each other. Microwaves make water molecules and some fat molecules spin faster. The more a molecule moves, the hotter it gets. That's how microwave ovens heat food. The microwaves can penetrate the food for about an inch. Then the heated food transfers heat to cooler areas. (That's why you have to rotate the food and stir it to distribute the heat evenly.)

This was loose popcorn popped in the microwave. Notice that it is covered with plastic wrap.

Corn Chips

Corn is not a snack food in Mexico. In the form of a baked cornmeal pancake known as a *tortilla*, it is a part of most daily meals. Corn chips are deep-fried pieces of tortillas.

The "dent" corn that is used for tortillas is related to the corn used for popcorn. The kernels are large, yellow or white, and have a dent in the top that formed when the seeds were dried. They are very different from the sweet corn we eat as a vegetable. They are starchy, not sweet, and they have a very tough hull that makes them almost impossible to grind into flour. So the Mexicans invented a clever way of removing the hull. They used the chemical *calcium oxide*, the main component of lime.

Lime is one of the world's oldest products of chemistry and still one of the most useful. It is a white powder that is used to harden concrete; it is mixed with sand and heated to form glass.

The masa flour in the bowl is used to make the tortillas that hold these nutritious taco sandwiches together. These same tortillas, when deep-fried, become corn chips.

Corn chips pile up as they come from the fryer. Corn-chip manufacturing is completely automated by machines: Conveyer belts move along the product.

A Chip of Another Color

When calcium oxide dissolves in water, it forms calcium hydroxide—a kind of solution called an *alkali*. Alkaline solutions have a slippery feel and can be quite corrosive; that is, they can burn the skin and eat through some materials. However, calcium hydroxide used for food is not strong enough to hurt you. Besides, most of it is washed away.

Most of the foods we eat are not alkaline. They are either chemically neutral or they are acidic. Acids react with alkalis (or bases) to neutralize them. The juice of a red cabbage can be used to test foods to see if they are acids, bases, or neutral. Like dent corn, red cabbage juice changes color when it comes into contact with different chemicals.

Chop about a cup of red cabbage, mix it in water, and stir. Strain the juice into a small jar. The purple color indicates a neutral food. Put a small amount of baking soda on a dish. Pour a little of the cabbage juice on it. The blue color indicates a base. Pour a little white vinegar or lemon juice in a small glass and add some cabbage juice. The pink color indicates an acid. Now mash some corn chips in a small dish and add a little water. When the chips have absorbed the water, pour a small amount of cabbage juice on them and stir. Wait as the color develops.

Test other foods around your kitchen. Most of them will be pink or purple—or, in other words, acidic or neutral.

Lime is made from limestone, seashells, or chalk by heating them in open air. The calcium in these substances combines with oxygen from the air to form calcium oxide.

Dent corn kernels are soaked overnight in water and lime. The hulls change color from pale yellow or white to dark yellow or orange as they react to the lime. Then they are cooked briefly. The kernels swell and soften as they absorb the water, and the partly dissolved hulls slough off. Next they are thoroughly washed to remove as much of the lime as possible. Now they are ready to be ground into a pasty dough called *masa* to become future tortillas or corn chips.

Nutritionists study food for its value in the human diet. There

Water and lime are added to dent corn kernels to make them swell and soften.

have been a lot of studies of corn, because it is the third most planted cereal crop after wheat and rice. Corn became a staple food for many poor people in rural areas because it was so easy to grow. But a diet of corn alone produced a terrible disease called *pellagra*. A pellagra victim suffered from skin rashes and sores, diarrhea, mental disorders, and eventually death. What could be the cause? The people in Mexico who ate mostly corn and beans never had this disease.

Science came to the rescue. It turns out that pellagra is caused by a lack of *niacin*, one of the B vitamins. Niacin is found in corn, but it can't be absorbed by humans unless it is treated by an alkali. So the lime-treated preparation of the Mexicans makes corn more nutritious. Beans also add another important nutrient—protein—to make a more balanced diet. Although corn may have traveled all over the world, unfortunately the healthful masa recipe did not travel with it.

Chocolate

Chocolate, like corn, is another gift from Mexico to the world. In the sixteenth century, when Spanish explorer Hernando Cortés met Montezuma, king of the Aztecs, he was served a bitter, hot drink called *chocolatl*, meaning "bitter water," in a golden goblet. For centuries, the people of South America and Mexico had been making hot drinks from the beans of the cacao plant. They thought so highly of the beverage that they used cacao beans as money. Cortés didn't much like the drink, but he brought the beans back to Spain, where someone thought the stuff would taste a lot better

The cacao beans in the freshly opened pod are a creamy white. The beans take on a chocolate color after a lot of processing.

if it were sweetened. This rare (and expensive) hot chocolate was so delicious that some people thought it was sinful.

The beans of the cacao plant have to go through a lot of processing before they become chocolate. In fact, it's amazing that anyone discovered it at all. The fruit is a green, football-shaped pod that turns golden or red as it ripens. It is split open, and the beans and pith are scooped out. Then the pith is allowed to rot, or *ferment,* as scientists would say. Fermenting warms up the beans so that some of the chocolate flavor starts to develop. Then the beans are cleaned, dried, roasted, and shelled.

The insides of the beans are called the *nibs*—and they are the essence of chocolate. The nibs are ground and warmed into a fluid paste called *chocolate liquor.* (*Liquor* means "liquid." Chocolate liquor contains no alcohol.) Unsweetened baking chocolate is a bar of cooled chocolate liquor.

Chocolate liquor contains two main parts—the dark brown, flavorful, chocolate-tasting solids and a white fat, called cocoa butter. Chocolate liquor is about

Baking chocolate isn't made to be eaten right out of the package. It is too bitter and crumbly.

The temperature of melted chocolate for candy manufacturing is carefully controlled by the central heating element in this mixer.

50 percent cocoa butter. In order to make chocolate candy, more cocoa butter is added to the chocolate liquor, making it 65 to 70 percent cocoa butter. To produce cocoa powder, the cocoa butter is removed from the liquor.

Cocoa butter has another property, which poses a challenge for candy makers. It forms very large fat crystals. In order to make smooth-textured candy, these crystals have to be made as small as possible, and they must be distributed evenly throughout the candy. This is done by heating, mixing, and cooling the melted chocolate in a process called *tempering*. Poorly tempered chocolate is crumbly, not smooth, tastes gritty, and has lots of white blotches of cocoa butter crystals on the surface. Well-tempered chocolate is smooth and glossy and breaks cleanly when you bend it.

One Cool Candy!

One of the most important properties of cocoa butter is the temperature at which it melts. It is a sharp melting point. This means that it changes from a solid to a liquid very quickly. Also this melting point—at around 93°F (34°C)—is slightly lower than the temperature of the human body. That's why it melts in your mouth. Put a square of chocolate candy on your tongue and hold it in your mouth as it melts. Resist the temptation to chew. It may get stuck to the roof of your mouth. Rub your tongue back and forth on the melting candy. Notice the cool feeling on your tongue and roof of your mouth. The melting chocolate is using heat from your body to melt, and as a result your skin feels cooler.

Blooming Chocolate

Chocolate is quite temperamental to work with. It's easy to see the bloom of cocoa butter crystals. Remove a chocolate bar from its wrapper and break it in half. Place one half of the unwrapped bar, molded side up, on a heat-resistant plate. Hold a hair dryer 3 inches (8 centimeters) above the surface of the candy bar, turn it on at high heat, and wave the dryer back and forth over the candy for one minute and fifteen seconds. Watch as the chocolate becomes shiny and the words start to disappear. You are melting the surface of the candy. Let the candy bar sit at room temperature for several hours (preferably overnight) to allow the chocolate to harden. Compare the heat-treated portion of the bar with the unheated bar. Notice the white spots on the surface and the bubbly, granular texture of the portion of the bar that is now out of temper.

Candy

The basic candy recipe is to dissolve sugar in water, cook it, and let it cool and harden. Pure water boils at 212°F (100°C). When you add sugar, the boiling point rises and water boils off, leaving the sugar behind. This makes the hot sugar solution more and more concentrated.

The temperature of the boiling syrup is a measure of how concentrated the solution is. The more water the syrup has in it, the softer the candy.

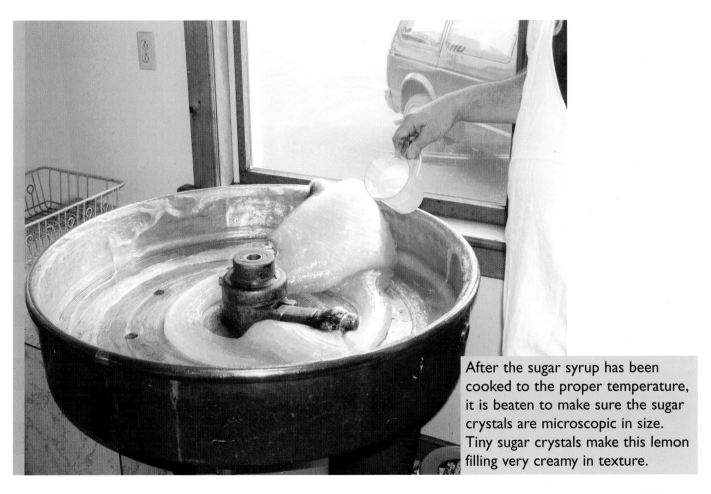

After the sugar syrup has been cooked to the proper temperature, it is beaten to make sure the sugar crystals are microscopic in size. Tiny sugar crystals make this lemon filling very creamy in texture.

Rock Candy

Rock candy is just slow-grown sugar crystals. To make it, you must have an adult present, as you will be using the stove. In a small saucepan dissolve 2 cups of sugar in 1 cup of water. Heat to the boiling point, stirring the mixture with a wooden spoon. When the sugar is dissolved, the solution will be clear. Add another $1/2$ cup of sugar while stirring. Let the solution cool slightly before pouring it into a glass jar. Cover the jar with a plastic coffee can lid. Stick a wooden skewer down through a hole in the center of the lid but don't let the skewer touch the bottom of the jar. Wait. It may take at least a week before sugar crystals form along the skewer (as well as the sides of the jar).

Notice the shape of the crystals. Use a magnifying lens to examine the shape of the crystals in a sugar bowl. Do they have the same shape? Crystals show the way the molecules fit into each other. The flat sides of the sugar crystals fit together like bricks. The shape of the large crystal is made possible by the shape of its building blocks.

Corn syrup is added to candy as a stabilizer. It keeps sugar crystals from forming.

When the temperature is 235°F (113°C), the syrup is 85 percent sugar, and a cook can make fudge. At 275°F (135°C), it is 90 percent sugar, and a cook can make taffy. When the temperature reaches 300°F (149°C), the mixture is almost pure sugar, and a cook can make hard candy.

Food chemists have a very good understanding of how the structure of sugar crystals determines the kind of candy that can be made. The simplest kind of crystalline candy is rock candy.

The syrup can be tested by a thermometer or by dropping it into cold water.

Potato Chips

Slice potatoes very thin and fry them in hot oil and you've got potato chips. The hot oil replaces up to 80 percent of the water in the potatoes, the slice is twice as thin as it started, and the result is crisp and crunchy, not soggy.

Browning occurs in potato chips as the starch caramelizes with heat. The brown color is due to the presence of the element carbon.

The "Birth" of the Potato Chip

Potato chips were invented by a chef in 1853 in Saratoga, New York. A restaurant guest requested very thin French fries. The chef cut the potatoes thinner than usual, but they were sent back. So the chef decided to cut them so thin that a fork couldn't spear them. The paper-thin, brown potato chips were a huge hit with the customer. They became a specialty of the restaurant and were known as "Saratoga chips." Today, potato chips are America's favorite snack food.

Two problems had to be solved in order to mass-produce potato chips to make them available to everyone. First, they had to be kept fresh. If potato chips are exposed to light and air, the oxygen reacts with the fat. This produces *rancidity*—many

These mass produced potato chips are coming from the fryer.

different chemicals that have unpleasant tastes and smells develop in the fat. Second, the chips are fragile and can break easily during shipping.

These problems are solved by packaging the chips in special bags lined with foil so that light can't get in. Manufacturers don't use clear plastic bags or even translucent materials that allow some light through. Also, air (which contains oxygen) is replaced with nitrogen gas. Nitrogen is a nonreactive gas, so it doesn't react with fat like oxygen does. It keeps the chips fresh. The nitrogen also cushions the chips against breakage.

At the end of the process, potato chips are enclosed in a light-tight foil bag filled with nitrogen.

Extinguish a Flame with a Bag of Potato Chips

Nitrogen is a nonreactive gas that makes up almost 80 percent of the air. A candle will not burn in pure nitrogen. See for yourself.

Ask an adult to light a candle for you. Blow gently on the flame. Your breath contains oxygen, and the gentle blowing fans the flame, it does not extinguish it. (If you blow hard, the force of the air's motion keeps the flame from getting oxygen, and it will go out.) With a scissors, cut off the corner of a bag of potato chips to make a tiny hole. Aim the hole at the flame and squeeze the bag gently so that you flood the flame with nitrogen. (You don't want to squeeze so hard that the force blows out the flame.) The nitrogen deprives the flame of oxygen from the air, and it will go out.

Making Rancid Potato Chips

Put some of the chips from your freshly opened bag in a glass bowl in the sunlight. Leave the rest of the chips in the bag and close it tightly with a bag clip or twist tie. Taste a chip from each group every day for a week. How do the flavors compare? Rancid potato chips won't harm you, but you'll get an insight into the problem facing potato chip manufacturers—keeping their product fresh as long as possible.

Soda

Has all this talk about junk food made you thirsty? Drink up as we talk about soda. A regular 12-ounce (36-milliliter) can of soda contains about 10 teaspoons of sugar, which weighs almost 1 1/2 ounce (42 grams). A diet soda is usually sweetened with aspartame—a chemical that is 160 to 200 times sweeter than sugar. Since only a pinch is needed to sweeten a diet drink, it should weigh less than a regular soda. You can easily check this out using a sink full of water. (See the "Lite" Detection sidebar on the next page for directions.)

The carbon dioxide gas in soda can be released explosively when you open a can.

"Lite" Detection

Hold a can of regular soda and another of diet soda under water in a sink filled to a depth of at least 8 inches (20 centimeters). Let go of the cans. Notice which can floats higher. It will always be the diet variety. The volumes of the cans are the same. But there is more matter dissolved in the regular soda, making it denser. This is such a reliable test that you can tell diet soda from regular soda even if you're blindfolded.

The Fire Extinguisher

Since it is heavier than air, the carbon dioxide released in a freshly poured glass of soda will sit on the top of the beverage for a while before drifting off into the air. It will also put out a fire. Here's a simple test. Ask an adult to strike a match and hold it just over the surface of the soda. Amazing, but a glass of soda can extinguish the flame.

The bubbles in soda are carbon dioxide gas—a gas that is heavier than oxygen and nitrogen, the two main gases in the air. Under pressure, carbon dioxide dissolves in soda. When you open a can or bottle of soda, you release this pressure and the bubbles quickly foam to the surface. Warming up the soda also makes the gas come out of the solution. So does shaking the can. Open a can of soda that's been rolling around the trunk of a car on a summer day, and try keeping your hands dry at the same time.

You Are What You Eat

Just like food, your body is made up of protein, fat, and carbohydrates. The food you eat is processed by your body so that you absorb it. Staying healthy and fit depends on what you eat and how much you eat.

All packaged foods have a label telling you what's in the food, as well as its nutritional value. Nutrition labeling for raw foods like fruits and vegetables is not required, but many providers do it anyway. There are labs all over the country that use scientific instruments to analyze foods. Do you know how to read a label on food? It's a lot of fun.

Most people don't pay attention to the portion size for potato chips. It's usually about twenty chips.

Nutrition Facts

Serving Size 1 ounce (28g/about 1 chips)
Servings Per Container About 11.5

Amount Per Serving

Calories 150   Calories from Fat 90

	% Daily Value*
Total Fat 10g	15%
Saturated Fat 3g	15%
Trans Fat 0g	
Cholesterol 0mg	0%
Sodium 310mg	13%
Total Carbohydrate 14g	5%
Dietary Fiber 1g	4%
Sugars 3g	
Protein 2g	

Vitamin A 0%	•	Vitamin C 10%
Calcium 0%	•	Iron 2%
Niacin 4%	•	Vitamin B_6 6%
Phosphorus 4%	•	Magnesium 4%

* Percent Daily Values are based on a 2,000 calorie diet. Your daily values may be higher or lower depending on your calorie needs:

		Calories: 2,000	2,500
Total Fat	Less than	65g	80g
Sat Fat	Less than	20g	25g
Cholesterol	Less than	300mg	300mg
Sodium	Less than	2,400mg	2,400mg
Total Carbohydrate		300g	375g
Dietary Fiber		25g	30g

Calories per gram:
Fat 9 • Carbohydrate 4 • Protein 4

INGREDIENTS: CHOICE POTATOES COOKED IN VEGETABLE OIL (CONTAINS ONE OR MORE OF THE FOLLOWING: CORN, COTTONSEED, SOYBEAN), SALT AND SEASONING ADDED.

SEASONING: WHEY, DEXTROSE, WHEAT FLOUR, NONFAT MILK, BUTTERMILK SOLIDS, SALT, CORN SYRUP SOLIDS, BUTTER SOLIDS, DEHYDRATED ONION, SUGAR, SOUR CREAM SOLIDS, MONOSODIUM GLUTAMATE, PARSLEY, CITRIC ACID, ARTIFICIAL FLAVOR, PARTIALLY HYDROGENATED VEGETABLE OIL (SOYBEAN, COTTONSEED), MALIC ACID, SODIUM DIACETATE, LACTOSE.

nutrition
facts

serving
size

amount per
serving

calories

calories
from fat

fats

saturated fats

unsaturated fats

cholesterol

sodium

carbohydrates

dietary fiber

proteins

iron

Nutrition Facts

Serving Size 3 tbsp (39 g) unpopped
(makes 6 cups popped)
Servings Per Bag about 2
Servings Per Box about 6

Amount Per Serving	3 tbsp Unpopped	6 cups (30 g) Popped	1 cup Popped
Calories	120	110	20
Calories from Fat	20	15	0
	% Daily Value		
Total Fat 2g	3%	3%	0%
Saturated Fat 0g	0%	0%	0%
Polyunsaturated Fat 1g			
Monounsaturated Fat 1g			
Cholesterol 0mg	0%	0%	0%
Sodium 380mg	16%	10%	2%
Total Carbohydrate 26g	9%	8%	1%
Dietary Fiber 4g	16%	15%	3%
Protein 4g			
Iron	4%	2%	0%

percentage
of the
daily value

g   mg

40

The main section of the label gives the **nutrition facts** for the particular package.

The **serving size** is very important. If you eat more than the serving size, you are getting more of all the other items measured.

The **amount per serving** heading tells how much of all the other listed measurements there are in the serving size stated at the top.

Food is fuel for your body. You burn it as you use it. **Calories** actually measure how much heat a food gives off when it is burned in a machine called a *calorimeter*. If you eat more calories than you need for your growth and all your activities, you will become overweight.

The government set 2,000 calories a day as a baseline for all the other numbers put on labels, such as the different **percentages of the daily value**. But this number of calories is only a guideline. The calories a person needs depends on how active and how big he or she is.

Calories from fat are not burned by your body as quickly as calories from carbohydrates and proteins. That's why people trying to lose weight are successful if they limit the number of calories from fat that they eat. Foods with more than 20 percent fat are not as healthy as foods with a low fat content.

Fats are greasy fats and oils. Extra, unneeded calories are stored as fat in your body. Some fat is important in your diet, but too much can be a problem.

Saturated fats are those that are solid at room temperature, such as butter.

Unsaturated fats, or oils, are healthier than the saturated kind.

Cholesterol is a fat-related material. It is connected to heart disease. So you want to limit the amount of cholesterol that you eat.

Sodium is a part of salt. Too much salt is bad for people who have high blood pressure. It also can make you thirsty.

Carbohydrates are sugars, starches, and fiber.

Most Americans don't eat enough **dietary fiber**, which is good for your digestion. Popcorn is high in fiber, so it is a healthier snack choice than many others.

Proteins are found in all living things. You are made mostly of protein, so it is also an essential part of your diet. You eat it in meat and fish and chicken. Vegetable protein is found in beans and tofu.

Iron is a metal that makes blood red. It is needed to help the blood carry oxygen to all the cells of your body.

G stands for grams. A gram is a measure of weight. It is as much as the weight of $1/5$ teaspoon of water. It is recommended that you don't eat more than 20 grams of saturated fat a day—that's only 4 teaspoons.

Mg means milligram. It is $1/10$ of a gram, a very small amount.

Sometimes **vitamins** and **calcium** are listed at the bottom of the chart. Vitamins are needed to help your body chemistry. If you don't have enough, you can get sick. Vitamin A helps us see in dim light and is needed to grow healthy hair, nails, and bones. It is found in fruits and vegetables. Vitamin C keeps our gums healthy and helps vitamin A do its job. Calcium is a mineral that is needed to help form strong bones.

What's It Worth to You?

Popcorn: A portion of popped corn is 4 cups. If you eat plain popped corn, without butter, it's a fairly healthy snack. It is only 80 calories and will give you almost one quarter of the daily value for fiber you need. Substitute butter-flavored popcorn, and a portion is 140 calories, with half of the calories coming from fat.

Corn chips: A portion is only about 20 chips and has 120 calories. Oil is the second ingredient listed, and it will give you about 9 percent of your daily value for fat.

Chocolate: A portion is one small (3-ounce or 84-gram) candy bar. It has 230 calories, with 120 calories from fat. It will give you 20 percent of your daily value for fat, but 45 percent of your daily value for saturated fat. Sugar is the second ingredient listed after milk chocolate.

Eat two of these and forget about eating any more fat for the day. You've just about filled your daily requirement.

Recent scientific studies have found that chocolate (these are wrapped in foil) contains chemicals that help cardiovascular health.

Candy: A portion of hard candy is one sour ball. It has only 20 calories and provides only 2 percent of the daily value for carbohydrates. Sugar is the ingredient listed first.

Potato chips: A serving size is about 20 chips, which is worth 150 calories. It provides about 15 percent of the daily value for fat. Oil is the second ingredient listed after potatoes.

Soda: A 12-ounce (36-milliliter) can of cola contains 140 calories, and the first ingredients listed are sugar and syrup. One can will give you 14 percent of the daily value for carbohydrates.

The best thing about knowing more about junk food is that you have the power to choose for yourself. Do you need to cut down on the junk food you eat? Are you eating a healthy, balanced diet? Are your parents? The answers to these questions are truly food for thought!

Key Words

To find out more, use your favorite search engine to look up the following terms on the Internet.

aspartame	dietary fiber
cacao plant	junk food
calories	microwaves
carbohydrates	nutrients
carbon dioxide	saturated fats
daily value	vitamins

Index

Page numbers in *italics* refer to illustrations.

About the Author

Ever since *Science Experiments You Can Eat*, Vicki Cobb has delighted two generations with her scientific and playful look at the world. In the "Where's the Science Here?" series she pays attention to areas kids know are FUN. Take junk food, for instance. She loved learning new ways to eat chocolate and put out candles with potato chip bags. She even loved learning how to read nutrition labels. Going to the supermarket now has a whole new meaning. Visit Vicki at: www.vickicobb.com.

About the Photographer

Michael Gold is a commercial photographer who has worked on assignment for some of the most exciting clients, including *The New York Times*, *Fortune*, *Esquire*, American Express, BMW, Mobil, *Opera News*, and many more. His work includes food, internationally known celebrities, advertising, products, fashion, and corporate photography. He has had nine one-man exhibitions, portfolios published in *Popular Photography Magazine* and *Camera 35 Magazine*, and is included in "Who Needs Parks?" and in *LIFE*'s first humor anthology, "LIFE Smiles Back."